W9-AAE-622

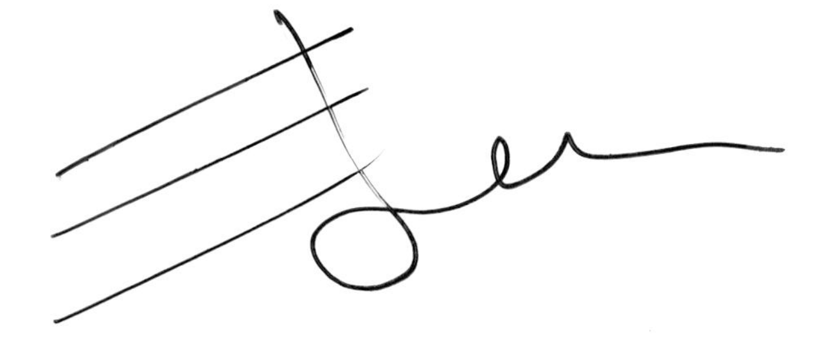

THE COCKTAIL PRIMER

WARNING:
DRINKING WHILE
PREGNANT CAN CAUSE
BIRTH DEFECTS
No Smoking
Please
MALIBU
PLYMOUTH
BOMBAY
DRY GIN
BEEFEATER
HENDRICK'S

ALL YOU NEED TO KNOW
TO MAKE THE PERFECT
DRINK

THE COCKTAILPRIMER

EBEN KLEMM
Master Mixologist

B.R. Guest Restaurants

Andrews McMeel
Publishing, LLC
Kansas City • Sydney • London

 For information, write Andrews McMeel Publishing, LLC, an Andrews McMeel Universal company, 1130 Walnut Street, Kansas City, Missouri 64106.

09 10 11 12 13 WKT 10 9 8 7 6 5 4 3 2 1

ISBN-13: 978-0-7407-7816-2
ISBN-10: 0-7407-7816-1

Library of Congress Control Number: 2009923690

www.andrewsmcmeel.com

Photographs by Miki Duisterhof
Book design by Michelle Farinella Design

CONTENTS

INTRODUCTION

Since the turn of the millennium, there has been a shift in how we think about crafting and, most important, consuming mixed drinks. What had been considered a mere beverage afterthought during the past three or four generations has now assumed as much creativity, inventiveness, and discipline as any other aspect of a good restaurant, bar, or nightclub. It is a given for most places to offer signature cocktails thoughtfully composed of fresh and exciting ingredients. No longer content with beer, wine, and vodka and tonics, increasingly sophisticated cocktail enthusiasts are demanding new and interesting drinks.

What this means is that when you hit the town, you have a far better chance of imbibing something delicious than anyone did fifty years ago. Even if you don't frequent bars and restaurants that serve a variety of cocktails, you are at least aware of the cocktail world celebrated in media and promoted heavily by liquor companies. Restaurants hire consultants to design drinks and train staff. Professional bartenders are expected to know how to make any specialty cocktail on the horizon and to stock a vast array of ingredients and garnishes at their bar. They are far more passionate about food, wine, beer, and service than their predecessors ever were. And their guests expect it.

Bartenders have approached the rediscovery of the classic American craft of making a great cocktail with amazing diversity. Some look directly at mixology as described in books from as far back as a hundred years ago; some look to local and seasonal ingredients and to the explosion in availability of new and formerly obscure ingredients to create fanciful drinks; and still others seize upon the culinary techniques developed by the world's most innovative chefs and translate them to the bar.

As the head bartender for B.R. Guest Restaurants, I develop drinks for a multitude of different restaurant concepts—from Dos Caminos Mexican restaurants, where we create margaritas by the thousands, to high-volume seafood and Asian restaurants, where the cocktails must reflect global and seasonal influences. I have had a chance to meet and work with some pioneers of modern mixology and to exchange ideas with the men and women who are as passionate about cocktails as I am; to visit significant bars and distilleries; and, of course, more important, to taste their magical liquid innovations. Yet what excites me most is to collaborate with the hundreds of men and women I work with every day. I learn alongside them not just what makes a good drink but what makes a good drink work.

this book

On the heels of this cocktail renaissance has come a veritable avalanche of cocktail books, as varied as the inspirations behind them. Any decent bookstore features a whole shelf of them, and every home furnishing store has a rack of them near the register.

So why on earth another one?

I've written and rewritten training manuals for our bartenders to establish both conceptwide and companywide consistency. I have turned to the dozens of cocktail books I have on my own shelf for help. All of them have a distinct point of view, and many are quite excellent, but I've found all but a few lacking.

Most cocktail books are encyclopedic, either in breadth (*A Thousand and One Cocktails*) or in depth (*Margaritas for Every Occasion*). Most bartenders will tell you that they really have to recall only a couple dozen cocktail recipes because that's all people ever order. While this may be true, it is difficult to introduce and train staff about new cocktails without at least some implicit knowledge of their origin. And if these encyclopedic texts are somewhat off-putting within a professional scenario, where almost every imaginable spirit and ingredient is available, they are doubly confusing, I imagine, for the home bartender. Which ingredients are important to keep on hand? Where do I start? Will I like it?

During my first year at B.R. Guest Restaurants, I tried to teach understandably skeptical bartenders the idea of drinks containing up to eight ingredients. The drinks were good, but a little difficult to learn and then reproduce consistently. The challenge was to make any drink I came up with fit within the concept of the twenty or so classic cocktails with which most people are comfortable.

A huge and thorough cocktail book usually arranges drinks either alphabetically or by spirit. Neither system does the aspiring cocktail enthusiast any good, because neither makes any attempt to tell you what the drink tastes like. Imagine a cookbook with recipes arranged from Avgolemono Soup to Zucchini Parmesan rather than by course. You'd have no idea where to begin. A book arranged by spirits assumes somehow that one ingredient defines the whole flavor profile of the cocktail. A martini and a gimlet both claim gin as their primary ingredient, but they do not taste at all alike and most likely would be enjoyed under very different circumstances. I enjoy a martini when I crave a cold and complex wave of alcohol washed over me, but I enjoy a gimlet when I need the refreshing spark of lime to soothe me after a hot day.

Inspired by one book, Gary Regan's *Joy of Mixology*, which is the first modern cocktail book to arrange drinks in families, I learned to think of all drinks, including specialty drinks, as descendants of very basic concepts. In the same way that in classic French cooking Nantua sauce, soubise, and Mornay are all children of béchamel, I will argue that the cosmopolitan, the margarita, and the sidecar all descend from a common concept and so are more related in flavor to one another than any of their individual ingredients might suggest. What I have discovered from teaching hundreds of classes is that cocktail recipes do not exist in a vacuum. Rather, almost every drink we can think to order or whip up at home is related to one of half a dozen master drinks. My goal in this book is to tell you to forget about the hundreds of drinks, and focus on the particulars. Understanding and appreciating these will propel you toward a mastery of any other drink you like.

First, we will put together a decent small bar, with the simplest equipment and the alcohol basics. Then, we will unite all our upcoming beverages with a discussion of drinking technology. I believe this to be the most important chapter in "getting" drinks, the place that separates the good one from the bad.

Finally, we will make drinks, divided into six classes: Each set of drinks is grouped by style and technique, and you will find this will help you organize them into what you like and when you like to serve them.

A perfect cocktail exists not in a recipe—for there are many different drinks for different moods—but in its execution. Despite the diversity we see on modern drink menus, cocktails remain combinations of but a few ingredients, blended by a choice of ever fewer techniques. It is easy for us to focus on the ingredients because they are the sexy components of our beverage, but how a drink is mixed is as important as what is mixed.

cocktails at home

Like many of my fellow drink inventors, I fancy myself a pretty good home cook. But while I suspect that, with enough time and concentration, I could make a pretty decent tortellini stuffed with three kinds of braised meat, I'd rather eat the dish at a good restaurant and instead prepare a nice roasted chicken at home. It is the same with the drinks I enjoy at home. While working, I love to create variations or develop complex infusions or foams because I have all the resources and equipment at my fingertips. But when I have a few friends over, I have much more important things to do than make anything but a perfect martini, a simple julep, or a seasonal fruit margarita, and I suspect you do, too.

I once had to reinterpret an old cocktail gadget for a hotel minibar project. The concept was to give the guests a calibrated cocktail mixer so that they could mix their own drinks from minibar contents; my task was to come up with enough unique cocktails from the minibar without having to resort to the peanuts. Operating out of a restaurant, with nearly unlimited access to ingredients, had not prepared me for this challenge, and I came to appreciate the notion that most of the time we have little but the most classic of cocktail flavors to work from. This is what this book builds on.

This book does not provide an unlimited repertoire of drinks, nor does it expose readers to many of the weirder combinations my bartenders and I try to put together in the restaurants. If you really love making drinks at home, I hope this lays the groundwork for learning how cocktails are assembled and how they're related so that you may not only have a greater appreciation for the wonderful explorations occurring in bars today but also feel comfortable creating your own unique cocktails and becoming your own personal mixologist.

CHAPTER 1
GETTING STARTED

All of the bars at my disposal at B.R. Guest Restaurants are professional bars. No matter what their aesthetic might be, they are designed for maximum efficiency and speed as well as the capacity to handle hundreds of different liquid combinations at a moment's request. They have ice wells, coolers, freezers, and custom-designed shelves, all constructed for the purpose of making this possible. A lot of assumptions about the ease of making cocktails are based on the very fact that they are being constructed in what is, essentially, a liquid factory.

An at-home bar is remarkably different from a licensed, commercial bar. The home bar is—well, how many of us actually have something that could be called a bar?—largely a different scenario altogether. It is a nice piece of furniture designed for two things: to hold the owner's classy-looking bottles and bar equipment for all to see and to maintain a safe separation between the mixer of drinks and those clamoring for one. It is not designed for making drinks as a professional bar is, and it never was intended to.

There is, in essence, an unwritten history of cocktails made at home and those made at a public bar. It's not unlike the differences between home cooking and restaurant cuisine. The home bartender labors under the weight of history, the breadth implied by old cocktail books, and the ease with which flawless cocktails are concocted in old movies. Hosts who believe they must be accountable at a party for a variety of different drinks find themselves sweating away, trapped behind the bar at their own party, a slave to elegance. And to make matters worse, earlier in the day they maxed out their credit cards buying far too many different bottles in the hope of mastering the craft on the spot.

If you read novels written during the last great cocktail era, or roughly the time between the two world wars of the twentieth century—novels by writers such as Dawn Powell and John O'Hara, verily, the Shakespeare and Flaubert of gin—it becomes apparent that when people came over "for drinks," there was really only one special cocktail served that night. Of course, there would be highballs and martinis, but in those halcyon days the home bar never collapsed under the weight of too many different drinks. Ideally, this is how you should go about entertaining, especially if, as for me, your bar is a kitchen countertop. Really, you have better things to do than make dozens of cocktails to order.

Interestingly enough, the professionally designed bar has remained largely unchanged for most of two centuries because, let's face it, it works perfectly for the barkeep. The ice lies in a large basin in front, with juices chilling on each side. There is a small drainage shelf for the bartender to assemble the drinks ordered; there is room to bring the bottles before the guest and to keep tools all in the same place. In short, the bar is designed to keep the bartender facing outward so that he can be involved in all of the action while being able to make a drink.

Compare this model of efficiency to an impromptu cocktail bar set up at home. The bottles are set up near the refrigerator or sink for ample access to ice or drainage. Either way, the drink maker's back is probably directed away from the guests. As the party heats up, and more bottles are opened and delivered, the small, concentrated bar space becomes cluttered with a forest of bottles, mashed citrus fruit, plastic wrappers, and melting ice cubes. The proper liquors become harder to find, the equipment less likely to be clean and at hand, and juices don't get replenished. Each successive cocktail-making expedition into this jungle becomes a little less committed to quality and a little more capable of increasing the collateral damage.

SETTING UP THE HOME BAR

If you want to offer your guests a complex range of drinks, here is how to set up the cocktail-making area in your own home to resemble a regular—professional—bar as much as possible.

1. Purchase liquor bottles no larger than 1 liter. The 1.75-liter bottles—or, as I like to call them, family-size—are too unwieldy for most uses and take up too much space.
2. Keep the ice near the work area and replenish it frequently. If you do not have an ice bucket, a large attractive bowl will suffice. If you entertain often, however, an ice bucket is a good investment and not a large one. (More on ice follows.)
3. Juices and premixed cocktails, should you need them, should be poured into clean glass bottles or pitchers. Keep them on a tray to reduce workspace moisture.
4. Keep fruit garnishes, which should be cut ahead of time, in a small bowl near the work area.
5. Religiously recap the liquor bottles immediately after use. You will be surprised by how quickly open bottles create messy spills all on their own. You can purchase attractive pour spouts for the bottles, which makes mixing drinks a little more fun and festive.
6. If you are working on a kitchen countertop or a table, arrange your supplies in an arc: Working clockwise, begin with the tray of juices. Follow it with the ice and then the liquor bottles. Put bowls of fruit garnishes in front of the ice, on the inside of the arc. Define the interior of the arc with clean kitchen towels and position a clean cutting board on top of them. This is where you will store your equipment and make your drinks. Make sure you have a place for garbage; a plastic bag or tub works best since cocktail-making refuse is wet.

7. Because you want to keep your equipment clean during service, set up bowls for dipping and rinsing if you are not near a sink. Having a sink in close proximity is ideal, which is why many homebound bartenders like to set up near a pantry sink or a small sink in a kitchen island. Of course, this will upset every good host's impossible goal of getting everyone out of the kitchen.
8. Unless you have a shelf immediately behind you or below you to store glassware, keep it to the outside of the bottle arc. This is not the most convenient place, but it keeps the glasses out of the way, safe and dry.

What you have put together is a space that, if not ideal for high-volume bartending, will keep the work area relatively uncluttered during a party so that you can focus your outward attention toward your guests. The discipline of the setup should help keep it neat, even when the party tilts in unanticipated directions.

Good bartending, like good cooking, occurs not merely when good ingredients meet a good recipe. Technique is crucial to success. Fortunately, compared with knife technique, bartending techniques are not that hard to develop. In the classes I teach for B.R. Guest bartenders, we spend almost as much time on this as on anything else.

TECHNIQUES FOR MAKING DRINKS

While I would never enter your house and criticize your bartending style, I would urge you to be consistent when mixing drinks. Follow the same steps each time you make a drink, no matter what it is. This way your hands and your arms will develop habits that anticipate your needs before you have to think about them. The best bartenders I know are not so much fast as efficient; they know exactly where everything is and what comes next. Although most home bartenders have no need for such highly developed skills, knowing how to do it right helps when it comes to mixing a stellar cocktail. The following techniques take you step by step through making a cocktail. Once these are mastered, the rest is easy.

Pouring

Hold liquor bottles by the neck. This might make them harder to lift, but you will have more control over volumes. Put the glass or cocktail shaker on the surface in front of you when you pour the drink. The motion of pouring should be swift and decisive. Lift the bottle up and flip the bottom in a quick, even arc. The more cautiously you pour, the more the bottle will betray you.

I heartily recommend using jiggers, at least at the onset of your career, so that you measure properly. It does not look amateurish; it looks like you care. When you fill the jigger, hold it over the vessel in which the drink will be made so that any extra liquid happily spills into the drink. Grasp it as its narrowest point, between two fingers, palm up, so that the slightest rotation of your fingers will pour the measure.

If a drink is made in the glass you are serving it in, as in a highball or simple shots on the rocks, fill the glass to the top with ice first and then build the drink over it. Do not serve drinks in glassware you have shaken! It is distinctly unsublime to present a glass whose outsides are sticky with fresh cocktail.

If a drink is mixed—in other words, shaken or stirred—add the ingredients to the smaller part of a Boston shaker, which is what most people think of as a cocktail shaker. (For more on this and other bar equipment, see page 19.) I recommend using a glass pint-size shaker for mixing drinks when you are learning so that you get a sense of what a pour size looks like. Add the ice last, once all of the liquids have been poured into the shaker. This way the ice

does not obscure your ability to recognize pour amounts, which you want to get a sense of as soon as you can. It also gives the ice less time to melt. Cold water doesn't help you out here.

Finally, return the bottles to their place as soon as you are finished pouring from them. This seemingly innocuous little habit is pounded into novice bartenders during their first week of training. It will help you develop more efficient bar skills than will any other technique I have described.

Shaking

Before you shake a cocktail, any cocktail, fill the shaker with ice to the brim once the ingredients for the drink are in the shaker cup. Shaking with ice accomplishes three things: it chills the drink to make it palatable; it adds a little water to it to make it refreshing; and it aerates the liquid to enhance its texture. Aeration occurs when the solid ice bashes around inside the shaker, and a full complement of ice is the only way to accomplish this effectively.

Secure the top of the shaker on the ice- and liquid-filled bottom, give it a small tap, and then wait a second or two to allow the shaker a little time to chill and contract to develop a seal. Now comes the fun part. While holding the top securely, lift up that shaker and SHAKE! By god, SHAKE! While you brandish it, hold the shaker firmly and remember that you are doing more than mixing the ingredients; you are emulsifying them with air and frigid water, establishing a fuller, richer, more evenly distributed flavor and mouth feel.

I recommend a number of hard, sharp shakes—up to twenty sometimes—to make a proper cocktail. In my recipes I state the number of shakes I would use when making the cocktail. Shake the shaker at least as many times as I suggest, but feel free to keep going for a few more shakes. It's an impressive, showy gesture that even veteran bartenders enjoy.

Shake in a way that is intense but not exhausting. A fixed wrist with big arm movement is the least efficient—and from personal experience, the most dangerous for those who might choose to walk behind you at a critical moment. Use as much wrist action as possible; it will reduce wild movements and make for a more efficient shake because it creates a more chaotic movement of ice inside the shaker.

If you doubt my proselytizing for shaking, ask yourself: Does the world of cocktails contain a more identifiable and pretty sound than the patter of ice in a shaker?

Stirring

There are times when a cocktail is more appropriately stirred than shaken, and those moments are described in chapters 2 and 3. There is something a little more elegant, and certainly a lot less frenetic and messy, about stirring. Both have their place, and one cannot be substituted for another.

If you've stored vodka in the freezer—which, by the way, is a terrible idea if you plan to mix drinks with the vodka—then you have observed how thick and almost syrupy alcohol becomes when chilled

to below freezing. It stands to reason, then, that alcohol can't mix by itself. Simply pouring various ingredients over ice is not enough to mix them without the addition of nonalcoholic ingredients or the help of a shaker. Stirring need not be as adamant an activity as shaking, but it must be done. A haphazard swirl of the shaker with one's wrist doesn't really cut it.

If there is a choice, I prefer to stir drinks in the metal, bottom half of the shaker since a cocktail stirred in glass takes longer to chill. Hold the bar spoon between your second and third fingers and dexterously rock it back and forth. Remember, you are making a martini, not bread dough. It is best to use a figure-eight stirring pattern at a deliberate but quick speed and to keep stirring until the outside of the vessel becomes visibly and tactilely chilled.

A frequent complaint of the halfhearted barkeep is that stirring takes up too much time. Nothing could be further from the truth. Consuming a poorly made drink, out of politeness, takes up too much time. If you, at home, feel stretched for time, simply stir your cocktail ten times and let it sit and chill while you complete some other tasks, but for no more than twenty seconds; then return and give it ten more stirs before serving. Somehow this seems classier, and, while I have no definitive proof, I feel it makes a tastier drink.

Some bartenders offer another argument against stirring, one that I find fatuous. They say it does not chill drinks as well as shaking them does. Their hypothesis, which claims as evidence the crystals of ice floating on the surface of a shaken drink, does not take into account the transfer of kinetic energy (heat) into the drink via that method. There really is no discernible difference in temperature between the two methods; they both accomplish the same thing: chilling the drinks by the transfer of heat, but with two very different physical concepts. We are left with aesthetics and flavor profiles on our barometer as to how we choose to blend our drink.

Serving

Half a decade ago it somehow became fashionable to fill a martini glass to the brim so that the customer had to take the first sip without lifting it off the bartop. With your cocktails, you are elevating your friends to the clouds of elegance, not converting them to horses at the trough. Nothing could be more unseemly; doing so also makes a mess and points to a bartender who is incapable of measuring drinks correctly. If, when making drinks at home, you end up with more drink than glass, wait until the recipient of the cocktail takes a sip or two before decanting the rest. Whether you pour the cocktails in front of your guests or serve them on a tray, you don't want your drinks to slosh around.

Shaken or stirred drinks that have been chilled with ice and then are meant to be served over ice, such as a margarita, should never be poured directly over the fresh ice in the glass in a way that causes some of the used ice to fall into the glass. Instead, the cocktail should be strained over the fresh ice. After all, the ice used to mix and chill the drink has broken apart, warmed up, and will not chill the drink as effectively as fresh ice; instead it will dilute it. Fresh ice keeps the drink colder longer and at the minimal level of dilution anticipated when we decide to make a particular cocktail.

THE ESSENTIALS OF A WELL-STOCKED BAR

For the sake of clarity, I will refer to liquors as either primary spirits or secondary spirits. Both have a place in the home bar.

Primary spirits are the foundation of a drink, the single monotone flavor around which the drink is wrapped. They are the key in which the symphony is composed. Primary spirits derive their flavor from what they were distilled from or flavors acquired through aging. That's it.

Secondary spirits are derived from primary spirits that have been sweetened or diluted or spiced—or all of the above. With their flavor profile ramped up to such an intense level, their role in the cocktail is, indeed, secondary. They lurk under the primary spirit and provide commentary. Rather than go too deeply into the history and complexity of these items, here are some basic descriptions and suggestions for stocking your bar.

The Primary Spirits

Gin: This is probably the one spirit without which we could not have cocktails as we know them. Gin, with its complex mélange of aromatics and spices, has tracked the evolution of the cocktail at every era and so is best equipped to fit any flavor profile of drink, from the sweet to the savory, from the fruity to the bitter. Why, then, do so many people have a hard time with gin? Because to shine, gin really needs to be part of a well-made cocktail. Too many people cannot get past having discovered gin in its least flattering form: by itself.

There is more diversity between the most similarly flavored brands of gin than between the two most diversely made vodkas, but if you must have one gin, make it a London-style gin. This is a sharp, heavily alcoholic, junipered beverage that we most associate with gin. After that, although I hate to recommend brands, you must have Plymouth, whose soft and delicate style makes it a winner with gin neophytes and for older, more simply styled cocktails. In recent years there has been a miniature explosion in artisanal gin makers, all of whom add their own innovations with new flavors and aromas. These gins are very often dedicated to heightening very particular, specific flavors, so they tend to be absolutely perfect in some cocktails and completely lose their identity in others.

Vodka: Mixologists tend to malign this category for its obvious lack of flavor and interest, but it is responsible for more of our salaries than we'd like to admit. We're still kind of right: vodka doesn't give anything to a drink but alcohol. It's happy to be invited to the party but is content to stand by the wall and look on, where it can still make an impression. Vodka is best when mixed with vivacious, fresh, and fruity ingredients so that those flavors come forth. And honestly, when you do that, you don't need anything expensive. Buy something smooth enough for you to enjoy on its own, and the rest will follow.

Rum: Nowhere does dollar for quality get you further than with rum. And good rum is really, really good. Rum's marketing problem is that it can be made anywhere in the world in any number of local styles, so it's rather difficult to figure out what your needs are. At the very least, you will need a white, unaged rum for the bright, citrusy, tropical cocktails and an aged, or dark, rum—not too aged, and certainly for no longer than ten years—for the darker, richer classics.

Tequila: Tequila is distilled from agave, a cactuslike plant related to the lily that grows well in many regions of Mexico. Perhaps the finest tequila is 100 percent blue agave tequila, which must be made from the Weber blue agave plant to rigorous specifications and only in certain Mexican states. Otherwise, tequila is available at different levels of aging and is so labeled: blanco, reposado, añejo, and extra añejo. Blanco, or silver, is unaged and is the best type for most tequila-based cocktails, such as margaritas. As tequilas age from reposado (two months to one year in oak) to añejo (one to three years in oak) to extra añejo (older than three years), they gradually lose the bright, fiery herbal and fruit flavors so wonderful for lime-based cocktails and obtain the spicy, oaky sophistication of scotch and brandy. In the world of cocktails, it is better to treat them as such.

When you buy tequila, avoid mixto, made when distillers cut the agave juice with other sugars, such as molasses or beet sugar. This doesn't always make for a bad-flavored bottle (although very often it does), but it won't give you tequila that is alive and brimming with the citrusy spiciness of real tequila. It's easy to avoid purchasing a mixto: look on the tequila's label for a statement that says 100 percent agave.

One more thing: A bottle of tequila will never contain a worm. A grub from the agave plant will sometimes appear in mezcal, a related spirit and one that I love but that is often too complex to mix into most simple cocktails.

Whiskey: For cocktails, it is far more important to have good-quality American whiskeys than to have whiskeys from anywhere else. American whiskeys—bourbon and rye—were around when the rules of cocktails were being defined, and so from an evolutionary point of view it stands to reason that they're the gold standard. Further, the fat, oaky opulence of American bourbons and ryes gives cocktails a substance that can't be found if you substitute the reedier, more intellectual scotches.

Scotch, on the other hand, hails from just one tiny region of the world, Scotland. Scotch whiskey is arguably the most diverse-tasting category of primary spirits out there. A cocktail book could not possibly hope to cover them all. To generalize grossly, most scotches either are single malts (100 percent barley-based spirits distilled and aged in one particular place) or blends (a mixture of single malts combined with other aged grain whiskeys). From a cocktail perspective, the relative sweetness and cost of blends make them good candidates for cocktails, while the diverse flavors and expense of malts recommend enjoyment on their own.

Brandy: The blessed union of oak and a grape spirit, brandy is probably the least used of the primary spirits in today's cocktails. And yet it does have a great history in the form. You can pretty much use it anywhere whiskey is called for. It is necessary for the sidecar, perhaps the most important cocktail of all time. You will have to wait a few chapters before I defend that statement, but in the meantime, one village-level brandy, or VS cognac, shall do fine for cocktail needs. Don't use anything nicer. The inventors of the drinks never did.

The Secondary Spirits

Wine-Based Spirits: We will talk of gin and vodka elsewhere, but now is as good a time as any to wonder why I do not give mixing liquors the same respect I give primary spirits. Vermouth is the salt and pepper of cocktails; even if you don't like the way vermouths taste, you should understand how they work. It is amazing to me how few Americans—even bartenders—actually know what vermouth tastes like. Part of the problem is that vermouth, because it is made from wine, starts to become stale not many days after being opened, and in these days of extremely dry martinis and half-

hearted Manhattans, we tend to keep the bottles open for far too long. My advice is to buy the 375-milliliter bottle and keep it in the refrigerator. You can always cook with the white vermouth.

Lillet, Dubonnet, and sherry also fall into this category of wine-based spirits and make minor, though significant, contributions to cocktails.

Fruit-Based Spirits: There is no end to this category; it really goes as far as the imagination takes you and is the most important category of secondary spirits in today's era of cocktails. For the home bar you need at least an orange-based liqueur (Cointreau or Triple Sec), an amaretto, a maraschino, and a cassis liqueur. These supply fruit flavors—real or artificial—as an undernote in complex sours. With cassis and all berry-flavored liqueurs, try to buy a good French brand instead of something artificially flavored. Per drink, you're not going to use a lot of it, so you can afford to get something good quality.

Herb-Based Spirits: The most mysterious of the categories, these are the spirits that go bump in your drink, and as they do, they throw down strange, complicated notes. Absinthe, anise, and licorice liqueurs come to mind, as well as chartreuse. The bitter spirits, such as Campari, fall into this category as well. Stocking a full quiver of these can be expensive and painstaking, but you need to have a few around which to build your cocktail expertise.

Sweet-Based Spirits: These include crème de cacao, cream liqueurs, and the like, sweet and syrupy liqueurs that have little prominence in the most basic cocktails. Still they can add unmistakable qualities to a more daring drink. Unless you are addicted to white Russians and similar drinks, you will find a little bit of one of these goes a long way.

THREE DIFFERENT BARS

If we take a look at how three different bars might be stocked, it's easy to see how a home-based bar might evolve as the bartender becomes more proficient and confident about making and serving cocktails. Each is different in terms of its alcohol and its aspirations.

Bar 1: "Hey, I just got a cocktail book!"

- London gin
- Grain vodka (made from grain rather than potatoes)
- White rum
- Blanco tequila
- Bourbon
- Sweet and dry vermouth
- Orange-based liqueur

This simple lineup, coupled with any mixers you choose to add, gives you the capacity to make just about any basic cocktail you want.

Bar 2: "I added these after reading the list for Bar 1!"

- Plymouth gin
- Aged rum
- Reposado tequila
- Rye
- Scotch
- Cognac
- Lillet
- Campari
- Maraschino
- Crème de cassis
- Amaretto

Most of the cocktails in this book can be made with this lineup, with enough extras left over to inspire you to branch out on your own.

Bar 3: "Uh-oh! Eben's coming over, so I had to stock up on some more!"

- New-style gins, such as Hendrick's and Aviation
- Potato vodka (vodka made from potatoes rather than grain)
- Aquavit
- Crème de cacao
- Crème de pêche
- Chartreuse
- Absinthe or Pernod
- Sake

This list is nearly arbitrary, but if you really want to spread your mixologist wings, these bottles are good to have on hand.

BAR ESSENTIALS: EQUIPMENT AND INGREDIENTS

Chances are if you are reading this book, you already own some cocktail-making equipment. If what you own is not what I recommend here, that is OK, but understand that the forthcoming recipes assume the use of the equipment I recommend. For those of you who have not yet rushed out to buy cocktail-making tools, I say: Although you may think other kitchen implements can be substituted for the equipment I suggest, they cannot. Drink making relies on such a large part on the fluid transition from step to step. Slotted spoons do not replace strainers, glassware does not stand in for a shaker, and the butt end of the Tabasco bottle does not substitute for a muddler.

The day of a party, squeeze a quart of fresh lime juice and a quart of fresh lemon juice. One quart will make approximately sixty drinks, which is not a bad way to start. A hand juicer—either the tabletop kind or the Latin American handheld type—is necessary. Professional bartenders are immune to the ravages of citrus juice, but if you have sensitive skin, you might want to wear latex gloves. Cocktails taste better when the citrus juice is strained of pulp. I recommend you make Simple Syrup (page 24) at the same time so that it has time to cool. What am I saying? You should always have simple syrup, so you should have made it days ago.

Equipment

I've spent my adult life working on bars, and I am a vocal proponent of not spending a lot of money on fancy bar equipment. This could be because I've seen too much get lost or stolen or just disappear.

After once gently setting an antique crystal martini pitcher containing a deliciously frosted cocktail on my kitchen counter and watching it vaporize instantly into a glass cloud, I have been a fan of the basics. Inexpensive bar equipment evolved for a reason: it remains the most efficient way to make cocktails.

Here Is What You Absolutely Need:

Shaker: Yes, you need one. I wholly endorse avoiding three- or four-piece shaker sets and opting for a simple Boston shaker from a local restaurant supply store. It has the simple two cups that fit over each other. The metal-on-metal set is a little more efficient for chilling drinks and makes a nicer shaking sound, depending on whether you prefer a heckita-heckita-heckita to a shooka-shooka-shooka, but the pint glass on metal is a bit better when you're getting started because you learn how much you are pouring. I am not the biggest fan of the three-piece cobbler shaker (vessel, built-in strainer, cap) for two reasons: the shattered ice tends to clog their built-in strainers, and, at least for the novice, it's tough to get the proper emulsion for a well-made cocktail. Save your money for the booze.

Strainers: You need two Hawthorne strainers (those are the cute guys with the spring as a seal) and one julep strainer. The former locks over the larger half of the shaker, and the julep strainer fits over the smaller half (how you choose to pour is a matter of personal style and bears great self-reflection). Looking like a giant spoon with holes in it, the julep strainer is better for straining drinks with chunks of fruit or herb in them. There are no springs in which ingredients can become snarled.

Bar Spoon: Because nothing undermines the elegance of a martini like stirring it with a soupspoon, you must have a bar spoon. The length of the spoon is there to give you intellectual distance from your cocktail and to remind you that you should be saying witty things to your friends rather than looking like you are trying to remember how to make the drink.

Muddler: A quality muddler is actually just about the trickiest piece of equipment to find. I am not sure why the world is flooded with muddlers that are (1) too small and insubstantial to muddle a drink efficiently, (2) painted so that colored flakes fall into the drink, or (3) made of wood that rots.

Be on the lookout for a muddler that is a rounded piece of hardwood or fruitwood, 8 to 10 inches long, with enough weight behind it so that a few strident pounds will do the job. There are also steel and glass ones that are nice; the key is heft. Don't fret too much; I am going to spend a lot of time trying to convince you to avoid muddling at home anyway.

Citrus Knife and Small Cutting Board: Any of these that you have for your general kitchen use will work fine.

Channel Knife and Fruit Peeler: It depends on how fancified you want your fruit twists to look, but you'll want one of these for your garnishes. Do you want to be wielding a knife after your second martini? Channel knives and fruit peelers are much safer.

Ice Scoop: You need one of these so that you are not the person who uses his hands to fill up glasses and shakers with ice cubes or, even worse, uses a shaker or a glass to do so, either contaminating the ice with the last cocktail made or possibly endangering the guests with glass shards.

Jigger Measures: Jiggers are absolutely necessary when you are getting started. They give a sense of what the correct measurement looks like. Today the best cocktail bars require even experienced staff to use them. It also helps to use a pint glass as half of your shaker set. Even if you don't intend to measure pours, it is best to train yourself to get a sense of what you are pouring.

Tools we would recognize as dedicated cocktail equipment have been around for about two centuries, but home cocktail equipment has been with us a less than a hundred years—although when you browse in some antique stores, it appears as if half of our GDP was committed to the manufacture of cocktail equipment. As you become comfortable with your technique, add some soul to your drink weapons by adding a few vintage jiggers, pour spouts, and the like.

Ingredients

Ice: You don't have enough. I'm serious. If you want to make quality cocktails in abundance, double what you think you need. A standard 5-pound bag of ice fills a Boston shaker only ten times. The ice from your average ice cube tray makes only one cocktail if you strain the drink over fresh ice, as I recommend.

Ice serves three purposes: to help mix the drink, to add water to it, and to cool it. The best ice cubes are fresh and large, directly from the freezer. Small cubes, ice that's been sitting out for a while, and ice that has been sitting in bags so long it has begun to crystallize all compromise the ice's ability to work.

Outside of a small party, most freezers lack the capacity to generate enough ice to keep up. And if you live, as I do, in a city where most freezers are about the size of a waffle iron, you are going to have to resort to commercial ice. It is best to buy it the day of your event and, if you can, buy bags where the ice has not consolidated into one big block. At any rate, break apart the ice (while still in the bag) when you get it home from the store, because this is the last thing you want to have to attempt later. Believe me.

Crushed Ice: There are certain drinks—such as muddled mint drinks—where the highest level of quality is achieved only when poured over crushed ice. If you do not have a freezer that makes crushed ice, you will have to crush it yourself. This can be time consuming and distracting during a party.

There are hand-cranked ice crushers, which generate a fairly consistent grade of crystal. Lewis bags—linen sacks that hold ice while you beat something against them and work out your inner demons—work, too. A clean kitchen towel will work as well, or even a plastic zipper bag, though the water generated has nowhere to go and the plastic can break. Well-crushed ice depends on the zeal with which you strike the ice and how capable you are of getting out the frustrations of your day while doing so.

Soda: This book is only minimally concerned with drinks that include soda as an ingredient, but I would not want to set up a little home bar without them. If worse comes to worst and you find yourself entirely in the weeds, you can distract the thirsty hordes with gin and tonic.

I recommend using 7- or 12-ounce bottles for several reasons: they are easier to handle, create less waste, look cute, and, most important, are easier to store cold. This is important because it increases and preserves their effervescence. When it comes to these beverages, I suggest stocking the basics: soda, tonic, and cola (consider augmenting a home bar with high-quality citrus soda, ginger beer, and an organic tonic water).

There are so many more small-brand, high-quality sodas available now than ever before, which can only lead to more easy drink ideas. My one caveat is to remember that perceptions of how a drink tastes are based on our memory of how it should taste. A gin and tonic made with tonic made from real quinine is far more refreshing and also remarkably less bitter than one made with the artificially flavored tonics.

Additional Liquids: One bottle of angostura bitters, 1 bottle of orange bitters, and one bottle of Rose's lime juice.

Simple syrup is also essential. You will find yourself using this easy-to-make sugar syrup over and over again as you craft cocktails. If you need more than a cup of syrup, the recipe is easy to double or triple—and simple syrup keeps for weeks in the refrigerator.

SIMPLE SYRUP

Makes about 8 ounces or 1 cup

1 cup water
1 cup granulated sugar

In a saucepan, bring the water and sugar to a boil over high heat. Reduce the heat to medium-high and simmer, stirring occasionally, until the sugar dissolves and the syrup is clear, about 5 minutes. Let the syrup cool and store in a tightly lidded container in the refrigerator for up to 1 month.

Garnishes: You'll want to have a good supply of fresh, basic garnishes, including olives, oranges, lemons, and limes. You'll also need some good-quality cocktail cherries, which you can easily make yourself. They are best made with sour cherries when they are in season. Sour cherries are sometimes called pie cherries and should not be confused with sweet, eating cherries. Here is my recipe:

COCKTAIL CHERRIES

Makes about 2 quarts

2 quarts sour red cherries, stemmed and pitted
1 cup grenadine
8 ounces (1 cup) vodka
8 ounces (1 cup) golden rum, brandy, or American whiskey, such as bourbon or rye
2 ounces (1/4 cup) maraschino liqueur

Divide the cherries between two 1-quart jars with tight-fitting lids. In a large glass measuring cup or similar container, stir together the grenadine, vodka, rum, and maraschino liqueur. Pour the mixture evenly over the cherries in both jars. Cover tightly and refrigerate for at least 1 week.

Metric Conversion Formulas

To Convert	Multiply
Teaspoons to milliliters	Teaspoons by 4.93
Tablespoons to milliliters	Tablespoons by 14.79
Fluid ounces to milliliters	Fluid ounces by 29.57
Cups to milliliters	Cups by 236.59
Cups to liters	Cups by .236
Pints to liters	Pints by .473
Quarts to liters	Quarts by .946
Gallons to liters	Gallons by 3.785

CHAPTER 2
MARTINI'S CHILDREN

Martini
Vesper
Negroni

Makeup: Clear spirits, no mixers

Complexity: High

Sweetness: Low

Acidity: Low

Strength: High

Level of Refreshment: Low

The most ethereal of mixed drinks, martinis and their kin are the cocktails to drink when we want to taste jazz. The flavors of delicate, herbaceous liqueurs are rendered subtle when diluted and underscored by a powerful white spirit, such as gin or vodka—truly a pillow of eiderdown wrapped around a brick. Even if other cocktails these days are more popular, or perhaps more *au courant*, I would be hard pressed to name a family of drinks that connotes better what the aesthetic vision of mixed drinks is all about. We drink these cocktails for their strength, their palate-preserving bittersweet song, and, most important, their clarity. There is something discreetly urbane about them: I can't imagine a night at the bar at Primehouse New York, our Gotham-inspired steakhouse, without seeing a bartender stirring up a martini or related cocktail.

Drinks in this family are those where a clear white spirit—most often gin or vodka—is mixed with light, delicate liqueurs and bitters. The primary spirit serves essentially as a backdrop for the secondary spirit, its mild flavor notes helping to exalt the other. These are strong cocktails, and without the fruit, juices, or sweet liqueurs found in other offerings, their flavors are more delicate and nuanced. It is because of this that they remain the best cocktails to consume if wine is to follow. And because the success of the cocktail is based on the interplay of what is usually no more than two or three ingredients, there is no other family of drinks that requires as much precise construction.

Since these cocktails are strong, we want them extremely cold. Because they are delicate, we do not want a surplus of—or any—water in the drinks. What is commonly referred to as bruising is nothing more complex than adding too much nonalcoholic liquid to the drink and thus unraveling the concentration of its flavors.

My friends love it when I order a martini at an unfamiliar bar. Years of making cocktails have trained me to be very particular about martinis. I get uptight and have no problem telling the bartender just to make me a gimlet. Now, I don't care how the bartender makes a margarita, but if I'm in the mood for a martini, I'll order it and it had better be good! I really hate to be that guy, you know the one, who spends a paragraph instructing the bartender how to make a proper one. More often than not—and this is what really delights my friends—I will be on the receiving end of a misguided lecture from the bartender on how the drink should be made, usually a version very far from what I like. This is when my companions expect me to smash something on the floor and shriek, "Don't you know who I am? DON'T YOU KNOW WHO I AM?"

Martinis and other drinks in that family are usually the best test of a bartender's skills. A good bartender knows that, with any of them, there is a dialogue with the guest, a dialogue of brands and garnishes, of wetness and dryness, a collaboration even, on constructing these most elegant drinks.

MAKING MARTINIS

These drinks should be stirred in a shaker that is completely filled with ice. Only then can you achieve the alcoholic balance that makes these cocktails great. It is a demonstration I perform in all my introductory classes for B.R. Guest bartenders: Make two gin martinis, shake one, stir the other, and try both. Not only does the shaken one look less appealing—an icy translucent soup next to a light-catching diamond—but its flavors are hard to discern. There is little truth that a shaken cocktail is colder than a stirred one; many are fooled by the presence of ice crystals in the shaken cocktail.

Some bartenders argue that stirring takes longer. Not really. Stirring cannot be duplicated with a gentle swirl of the shaker or by pouring the ingredients together and hoping that the gods of Brownian motion will do their part. When they hit ice, liquors become viscous at different rates; they will not mix on their own. When you stir a martini, you may be tempted to pause: fetch your glasses, prepare your garnishes, or take your hors d'oeuvres from the oven. I believe, on the basis of nothing but instinct, that a pause in the midst of stirring a drink somehow ties it together. Perhaps it's just the increased anticipation. (More on shaking and stirring can be found in Chapter 1.)

MARTINEZ

Serves 1

The prevailing wisdom is that this cocktail is not a predecessor of the martini. I am not sure if this is so, but it is a great old drink that deftly illustrates how complicated the flavors in this family can be. Upon tasting, you will be hard put to discern any particular flavor components here; the gin and the mixing spirits surrender their individual flavors to a sea of subtle spice.

2 ounces gin (I suggest London-style or Plymouth gin)
1 ounce sweet vermouth
1/4 ounce maraschino liqueur
Dash of orange bitters
Lemon peel, for garnish

Pour the gin, vermouth, maraschino liqueur, and bitters into an empty cocktail shaker. Fill the shaker completely with ice and stir with a bar spoon until the outside is cold. Strain and serve, garnished with lemon peel.

OLD-TIME MARTINI

Serves 1

The martini has become more popular than is good for it. Over the last century it has been reformulated so many times that it is barely recognizable as what it once was. Partly as a response to the American palate, it has become more savory, since more sweet drinks are available today. Mostly, however, improvements in the quality of gin and vodka have contributed to this trend. Less vermouth and less bitters are needed to obscure the power of bad booze, and the martini has metamorphosed from a dainty little winey cocktail to the powerhouse of alcohol it has become today.

1 1/2 ounces gin
1 1/2 ounces sweet vermouth
Dash of angostura bitters
Lemon peel, for garnish

Pour the gin, vermouth, and bitters into an empty cocktail shaker. Fill the shaker with ice and stir with a bar spoon until the outside is cold. Strain and serve, garnished with lemon peel.

1930S DRY MARTINI

Serves 1

Originally a dry martini referred not to the ratio of vermouth to gin but to the use of dry rather than sweet vermouth. Once this shift obliterated the original recipe, the ratio levels began to change. It seems we've gone up one level of reduction of vermouth every decade since the thirties.

2 ounces gin
1 ounce dry vermouth
Lemon peel or 3 small green olives, for garnish

Pour the gin and vermouth into an empty cocktail shaker. Fill the shaker with ice and stir with a bar spoon until the outside is cold. Strain and serve, garnished with lemon peel or olives.

NEW-TIME DRY MARTINI

Serves 1

I call this a new-time martini because it is made with vodka, which, rather than gin, seems to be the spirit of choice in the twenty-first century. As with a gin martini, the vermouth is important, although not everyone who makes a vodka martini agrees, and so many are devoid of vermouth. I garnish this with green olives, but if you substitute black olives, the drink becomes a buckeye.

3 ounces vodka or gin
Dash of dry vermouth
Lemon peel or 3 small green olives, for garnish

Pour the vodka and vermouth into an empty cocktail shaker. Fill the shaker with ice and stir with a bar spoon until the outside is cold. Strain and serve, garnished with lemon peel or olives.

Vodka Martinis

At B.R. Guest Restaurants and many others, we assume customers who order a vodka martini want no vermouth whatsoever. We don't stir these drinks; we shake the hell out of them on the assumption that the guest doesn't really want a cocktail but instead a double shot of ice-cold vodka. By doing so, we are adding more water to the drink and quietly rehydrating our friends, which they will thank us for later. Vodka has no flavor, and so it is impossible to bruise it or dilute it too much.

VESPER

Serves 1

This is a cocktail invented for Ian Fleming, the author of the James Bond series of thrillers. Historically it is supposed to be shaken, but I personally feel that this directive is more out of respect for Mr. Fleming's most famous creation than to the quality of the drink. The Vesper is something lovely. It is one of the few enduring creations to emerge from the 1960s and is a wonderful introduction to the martini family of drinks. The vodka keeps the gin in check, for those who are afraid of it, and the use of Lillet instead of vermouth lends it an exquisite orangey flavor.

2 ounces vodka
1 ounce gin
1/2 ounce Lillet Blonde
Lemon peel, for garnish

Pour the vodka, gin, and Lillet into an empty cocktail shaker. Fill the shaker with ice and stir with a bar spoon until the outside is cold. Strain and serve, garnished with lemon peel.

NEGRONI

Serves 1

There is no other drink that more surely defines a connoisseur. Conversely, there are fewer bartenders than there should be who know how to make this drink, which is a shame, as it is perhaps the best before-dinner cocktail ever invented. The bitters and the vermouth imply, rather than enforce, a spicy, citrusy flavor that opens the palate wide. This is the drink that made me want to bartend; I suspect in this I'm not alone.

2 ounces gin
1/2 ounce sweet vermouth
1/2 ounce Campari
Burnt orange peel, for garnish (see opposite page)

Pour the gin, vermouth, and Campari into an empty cocktail shaker. Fill the shaker with ice and stir with a bar spoon until the outside is cold. Strain and serve, garnished with burnt orange peel.

To Burn an Orange Peel

Burning an orange peel is a show-offy technique that adds the flavor of burned citrus oil to a drink and imparts a pleasant aromatic note with a slight orange nuance.

Cut a thumb-size slice of fresh orange peel, retaining as much of the white pith as possible. Strike a match and hold it in one hand. Hold the shorter end of the orange peel between the thumb and forefinger of the other hand and gently brush the flame against the orange peel.

Hold the still-lit match (or another one, if the first one is about to singe you) about 2 inches above the filled cocktail glass and hold the peel a similar distance over the match. Pinch the peel, releasing the oils. If you've done this correctly, the spray will ignite in a rather satisfying burst. Sometimes there is even a nifty little whoosh.

CHAPTER 3
BEYOND MANHATTANS

Makeup: Aged spirits, no mixers

Complexity: High

Sweetness: Medium

Acidity: Low

Strength: High

Level of Refreshment: Low

These are the dark, mystical cousins to the drinks of the previous chapter. Hailing from roughly the same time period, and made in the same fashion, these drinks layer the sweet intensity of barrel-aged spirits on top of the delicate flavors and fragrances of flavoring liquors. Glowing golden brown in the candle-light, the oak-sweetened liquors from which these potent drinks are crafted warm the body and mind in spite of the frigid temperatures at which they are served. If martinis and their ilk are serious and elegant, these are a little rowdier—but never out of control. Despite the similarities between martinis and these drinks, the flavor difference is significant enough to warrant a separate chapter expressly for them.

You may have heard bartenders divide the primary spirits into "whites" and "browns." All spirits emerge from the still clear, sugarless, and fiery. The delicate flavors that survive the rigors of several boilings are diluted, bottled, and sold. These are the white liquors: clean and vigorous vodka, gin, tequila blancos, and white rums. When a distilled spirit is allowed to age in oak for months or years, it changes radically. The time spent in these barrels depends on both the desired affect and the climate, but the result is always that the heretofore clear spirit turns the color of the wood even as it takes on its flavor. The flavor can be described roughly as sweet caramel, coconut, and winter spices (cinnamon, cloves, mace).

Beverage professionals toss around the word *oaky* a little too easily to describe flavors. I believe the term describes the sometimes subtle, sometimes bold potpourri of flavor notes of vanilla, butter, caramel, dill, pineapple, coconut, winter spices, and smoke that can be tasted at various intensities in all oak-aged wines and spirits. If you know the difference in taste between a steely Mâconnais or Chablis and a rich, buttery California Chardonnay, you have some idea of the effects of oak aging. The oak blesses the spirit with its rich brown hue, and while it rests in the barrels, the spirit absorbs the sap of the wood and sweetens. Think of oak aging as a mellowing process and the reduction of the fiery spiciness of raw liquor in exchange for smoothness and sweetness. Some spirits will never be anything but white (gin and vodka), some are both (rum and tequila), and others must be oaked even to be considered in their category (whiskey and brandy).

In one of my various and sometimes crackpot attempts to organize a taxonomy of the world of cocktails, I decided that all cocktails, no matter their style or family, could be divided into those whose flavors could be described as complementary and those whose flavors are supplementary. The ingredients in those described by the first category combine to form a concoction that tastes very like those ingredients. For example, a cosmopolitan tastes like citrus vodka and cranberry and lime juices; a vesper tastes like what you'd get if you mixed gin and Lillet. There is nothing wrong with this, of course, because those are nice flavors to have in your mouth.

In supplementary cocktails, something transcendent occurs, and the result is a flavor profile that can't quite be predicted from the recipe. With this family of cocktails, in myriad interactions between the sweet complexities of the brown spirits and the spicy or winey herbaceousness of their supporting casts, we begin to see a form. Given the splendor of this taste discovery, and the relative age of some of the drinks, I can't help wondering if the cocktails' earliest inventors raised their creations to their lips and suspected that they might have come across something truly special.

We want to make these drinks exactly like the ones in the martini family. Texture is almost as important to these cocktails as taste is. We do not want any excess water diluting the beautiful unctuousness of aged whiskies and brandies. One of the astonishing things about these drinks is how syrupy their consistency becomes when they are chilled deeply.

I could not end this chapter without addressing Canadian whiskey, and so I will come right out and say it: Canadian whiskey is a little like Canadian music. You can find better stuff below the forty-eighth parallel. During Prohibition in the United States, a lot of Canadian whiskey was snuck into America as a replacement for rye. Now that the United States is once again distilling plenty of its own stuff, I like to support it.

MANHATTAN

Serves 1

Originally made with rye, the Manhattan is equally good with bourbon. This recipe offers an acceptable ratio you can play with depending on your taste and the richness of the whiskey. Many neophyte bartenders believe that the whiskey-to-vermouth ratio compares with the gin-to-vermouth ratio for a martini and so splash just a little sweet vermouth into the shaker. This makes an acceptable drink, but it does not flavor the whiskey with enough intensity to make a proper Manhattan. In this case, these two most famous sons of the 1870s—Manhattans and martinis—have little in common.

2 ounces American whiskey, such as bourbon or rye
1 ounce sweet vermouth
2 dashes of angostura or orange bitters
Lemon peel, Cocktail Cherry (page 25), or maraschino cherry, for garnish (optional)

Fill a cocktail shaker with ice. Pour the whiskey and vermouth over the ice and then add a dash or two of bitters. Stir with a bar spoon until the outside of the shaker is cold. Strain the cocktail into a martini glass and garnish with lemon peel or a cherry.

Bourbon Versus Rye

No serious practitioner of drink making should be without America's two greatest spirits (I am frequently without either because I like them all too much). Set aside a night when you can, once and for all, determine which one you favor in your Manhattan. Because this isn't meant to be an exhaustive history of alcohol and cocktails, let me be as elementary as possible on their distinctions.

Bourbon is a whiskey whose mash bill—the term that describes the percentages of the different cereals used to make the beer that makes the whiskey—is at least 51 percent corn and must be aged at least two years in new, charred American white oak barrels. There is no need for the bourbon to be made in Kentucky or, for that matter, in its famed Bourbon County.

Rye replaces corn for rye whiskey. This spirit has been made longer than bourbon and was manufactured originally in relatively populated states such as Pennsylvania, Maryland, and Virginia. Its urban roots suggest that it, not its more popular brother, is the inspiration for most of the whiskey cocktails we venerate. Prohibition sent rye the way of the whooping crane. During those years, bootlegged Canadian whiskey replaced the demand for rye in the northern states and nearly wiped it out. Bourbon remained safe in the South. One of the many things we have to thank for the resurgent popularity of cocktails is that we are witnessing the comeback of diverse and very good rye whiskeys, although nearly all ryes are now made in bourbon distilleries.

Because of the appellations that determine how these spirits are made, bourbon tends to be sweeter, fruitier, and richer than rye. Rye aficionados appreciate it for its dark, dry spiciness. At long last, we are in the full throes of an American whiskey renaissance, and within their profiles, both styles offer tremendous variety of flavor and the possibility of many great cocktails.

SAZERAC

Serves 1

Now that absinthe is legal again (oh, that I could begin all my sentences like that!) and Peychaud's bitters is easier to obtain, we should begin to see this great old drink gain a foothold in cities not named New Orleans. I'm not sure if you can, scientifically speaking, pack more alcohol per volume into a drink. The absinthe is used just to coat the glass, which is an ideal way to control the amount of absinthe in the drink—and at the same time have the perfect measure.

1/2 ounce absinthe
1 1/2 ounces rye
1 1/2 ounces cognac
1/2 ounce Simple Syrup (page 24)
Dash of Peychaud's bitters
Lemon peel, for garnish

Fill a tumbler with crushed ice and pour the absinthe over it. Fill a cocktail shaker with ice. Pour the rye, cognac, and syrup over the ice in a shaker and then add a dash bitters. Stir with a bar spoon until the outside of the shaker is cold.

Discard the absinthe and ice from the tumbler. Strain the cocktail into the tumbler, garnish with the lemon peel, and serve.

LA DOLCE VITA

Serves 1

I created this simple little number for Vento, an Italian trattoria. I am always looking for ways to introduce people to whiskey cocktails. Sometimes the trick lies in using a sweet liquor to lure people in.

2 to 3 tablespoons powdered raspberry (see Note)
1 lemon wedge
2 ounces bourbon
1/2 ounce Carpano Antica
1/2 ounce Lazzaroni amaretto

Spread the fruit powder on a flat plate. Moisten the rim of a martini glass with the lemon wedge and upend the glass in the powder to coat the rim. Turn right side up.

Fill a cocktail shaker with ice. Pour the bourbon, Antica, and amaretto over the ice. Stir with a bar spoon until the outside of the shaker is cold. Strain the cocktail into the rimmed martini glass and serve.

NOTE: Powdered raspberry is available at specialty pastry stores. Fresh pomegranate seeds make a classy alternative.

EL DUCQUE

Serves 1

Since this category is so wed to the cocktails of the nineteenth century, it is sometimes difficult to translate the principles for what is relatively a twentieth-century ingredient. In this case, I am referring, of course, to tequila. Because tequila is wedded so closely to the margarita, we tend to overlook that, in its aged form, it can be utilized in this kind of cocktail. I created this cocktail for Mexican restaurant Dos Caminos because I love the deep, sweet herbaceousness of both añejo tequila and sweet sherry and thought they would work well together.

1 (6-inch) fresh pineapple wedge
About 1/4 cup demerara sugar
Seeds scraped from 1 vanilla bean
2 ounces añejo tequila
1 ounce Pedro Ximénez dessert sherry

Dust one side of the pineapple wedge with sugar and sprinkle with the vanilla seeds. Using a crème brûlée torch, caramelize the sugar. Alternatively, slide the pineapple under a broiler for 50 to 60 seconds to caramelize the sugar. Watch carefully that the pineapple does not burn.

Fill a cocktail shaker with ice. Pour the tequila and sherry over the ice. Stir with a bar spoon until the outside of the shaker is cold. Strain the cocktail into a martini glass and garnish with the pineapple wedge.

CHAPTER 4

THE SIMPLE SOURS

Makeup: One spirit plus acid

Complexity: Low

Sweetness: Medium

Acidity: High

Strength: Medium

Level of Refreshment: Medium to high

Although the drinks discussed in the first two chapters of this book stand tall for their classiness, complexity, and aesthetics, as well as for their extreme cocktailness, they lack one thing: refreshment. Those cocktails are all alcohol, and because of that their strength pigeonholes their consumption for a certain place and time. With sours, it's another story. They are all about rejuvenation and cool restoration.

Imagine it is the end of a hot July day and you are lucky enough to be reclining on a white wicker chaise on a breezy porch overlooking an endless stretch of blue water. Manhattans, anyone? I don't think so! Or how about conjuring up 2:00 A.M., time for one more round of drinks before your friends leave and pursue their own dark devices. You don't want to slap them over the head with martinis and send them out the door.

Simple sours are far more appropriate for these situations—and many more. They compromise the strength of a single spirit with the blessed dilution of citrus juice. The alcohol level drops, the water content increases, and the indescribable nuances derived from intermingling two or more alcohols are gone. In their place is the bright spark of fresh lemon or lime. Delicate sips become hearty quaffs, and the sweat on our brow isn't so heavy anymore.

THE ORIGIN OF THE SOUR

Simple sours are among the earliest of mixed drinks. They are an Occam's razor of beverages, embodying the principle that "entities should not be multiplied beyond necessity," as they crawl out of the primordial cocktail mud of naval punches and achieve definition in the very earliest of cocktail recipe books.

It is easy enough to divine their maritime roots, of course. Citrus was a staple of sailing ships, its generous vitamin C content necessary to ward off scurvy. Rum and gin were always on board, too, and so why not consume them both at the same time?

The explosion of the American railway system and the ability to ship Florida fruit to northern cities in the first half of the nineteenth century made citrus fruit a staple of cocktail bars. Perhaps patrons could fool themselves and their teetotaling friends that they were drinking lemonade when their glass was filled with fruit juice sweetened by simple syrup and fortified with only a "splash" of gin. Or perhaps a glass of freshly squeezed citrus juice seemed exotic when compared to a big old glass of whiskey. Most likely, with the population explosion and industrialization of the North American city, we just needed a cocktail that would beat the heat a little bit better.

RULES TO LIVE BY

No one writes about these cocktails as much as we ought to. They are overshadowed by their more complex descendants described in Chapter 5, and even I am a little bit challenged when it comes to simple sours. How much can you say about the marriage of one liquor and one juice, except that it's a happy union? But the beauty of these drinks lies in their very simplicity, as well as the relative ease of making them at home. Also, they are perfect cocktails on which to build and add to your bag of tricks and will help every amateur barkeep get a handle on everything that comes next.

Even though we add juice to the glass when we make sours, keep in mind that alcohol remains the primary ingredient. These are not screwdrivers or rusty nails. The cocktails I define as simple sours have an essential ratio of two parts liquor to one part mixer. The challenge is to blunt the alcohol by fully enveloping it with good-quality lemon or lime juice. There are two important parts of this:

1. **Use Fresh Juice.** The professional cocktail world drones on endlessly about fresh juice. I do not think you can make a proper cocktail unless the citrus juice has been squeezed within twenty-four hours of serving it, although even that is a little bit long. It's best to squeeze the juice as close to serving it as possible. Storage is sometimes a challenge in home bars, but try to make fresh fruit a priority. Calculate one lemon or lime for every two drinks—and then squeeze a dozen more! Squeezing by hand with a tabletop citrus press or a handheld squeezer, both of which will extract some of the flavorful citrus oils from the rind, increases the quality of the juice. Once the juice is squeezed, strain out the pulp. As much as you might think pulp will show your guests how fresh the juice is, a sour's taste is determined by the perfection of its mixing, which cannot be achieved with flavorless solids floating on its surface. Don't rely on your cocktail shaker to strain out the pulp after the drink is mixed. The pulp will clog the cocktail strainer and create a mess.

2. **Shake the Sour.** One of the first things I do in my bartending classes is to serve each new bartender a sour that has been shaken only halfheartedly and another that has had the living daylights shaken out of it! For good measure, I assume the bored and vacant expression of an uncaring barkeep during the former and the passionate and caring one (the barkeep I expect them to become) during the latter. When they taste the two cocktails side by side, they recognize how superior the well-shaken drink is. It's really kind of like night and day.

   An alcohol-based liquid (the spirit) and a water-based liquid (the fruit juice) are, from a basic chemical structure, two very different beasts. A stir or a gentle shake will not combine them correctly. A cocktail's tastiness lies, for the most part, in the efficacy of its blending, which signals the disappearance of the parts into the sum total. A poorly mixed sour will taste like its ingredients and no more. A poorly made gimlet, for example, will reveal itself as a limpid green series of flavors—sharp gin, sharp lime, sharp sugar—whereas a well-shaken one emerges as something wholly different: a fluffy white emulsion where the gin and lime are cooled and rounded by a perfect pillow of ice water. Try this for yourself if you don't believe me. For all cocktails with a juice component, the drink must be shaken and shaken well. (For more on shaking, see page 6.)

The ratios in the cocktails that follow might be too sweet or dry for you. I wholly recommend you adjust the ratio of your sour mix to suit your palate. I won't mind; really I won't.

In the age when sours were invented, cocktail garnishes tended to be on the baroque side, and drinks were often stacked high with citrus twists, grapes, and meringue. I use a simple cherry or thin round of orange, a garnish called a flag, but feel free to turn back the hands of time and use a fruit that underscores the flavors in the drink. One of the nice aspects of so simple a cocktail is that you can use it as a backdrop for other flavors such as flavored fruit schnapps.

TRUE SOUR

Serves 1

The following formula will work nearly universally for your favorite base spirit, provided it is not ghastly or too sweet. If you favor sweet liquors, you might want to skip the simple syrup and increase the amount of lemon juice.

2 ounces gin, whiskey (any kind), rum, amaretto, vodka, or eau-de-vie
1/2 ounce fresh lemon juice
1/2 ounce Simple Syrup (page 24)
Dash of angostura or orange bitters or splash of club soda
Maraschino cherry, orange round, or seasonal berries, for garnish

Pour the liquor, lemon juice, and simple syrup into a cocktail shaker. Fill the shaker with ice, cover, and shake vigorously 20 times.

Fill a rocks glass with fresh ice and strain the sour over the ice. Add a dash of bitters or a splash of soda and garnish with a cherry, orange round, or berries. The soda makes the drink a little livelier.

GIMLET

Serves 1

Few drinks inspire more disagreement about its proper constitution than the gimlet; the question is whether to use Rose's lime juice instead of fresh lime juice. There is a historical precedent for the bottled lime juice, and for a good part of the last fifty years Rose's was the closest you'd come to lime juice of any kind at any bar. If the 1950s are your aesthetic, there is a kind of cool "better living through chemistry" kind of vibe to the syrupy texture of Rose's. When you survey the crowded refrigerator door shelves in your neighbors' houses, you are likely to spy the familiar bottle of pale green syrup lurking there.

There are those who feel you should use fresh juice if at all possible, despite the cocktail's legacy. To be sure, there's a summeriness about fresh lime in this drink—especially if you make the gimlet with gin—that cannot be duplicated in any other drink.

I provide three ways to make the gimlet. The third is my middle-of-the-road version, and, like all middle-of-the-road things, it pleases everyone without commitment. It's quite nice, actually, especially if you've run out of simple syrup.

2 ounces gin or vodka
1 ounce Rose's lime juice
or
2 ounces gin or vodka
1/2 ounce fresh lime juice
1/2 ounce Simple Syrup (page 24)
or
2 ounces gin or vodka
1/2 ounce Rose's lime juice
1/2 ounce fresh lime juice
and (for all variations)
Lime wedge or grated lime zest, for garnish

Pour the gin or vodka, lime juice, and simple syrup (if using) into a cocktail shaker. Fill the shaker with ice, cover, and shake vigorously 20 times.

Strain the chilled and blended cocktail into a martini glass and serve straight up, garnished with the lime wedge or a pinch of grated lime zest floated on the surface.

NOTE: If you switch to white rum and use the lime juice and simple syrup, you will have a daiquiri. I don't like to hide this drink in the shadow of another, but it's clear to see that they are essentially the same drink. The daiquiri carries with it the same curse as the martini and the margarita: with great popularity come multiple variations, and with multiple variations come ideas that often lie very far from the drinks' original intent.

FIZZ

Serves 1

Egg whites are one of the best things you can introduce to alcohol and lemon. The resulting meringue settles on top of the cocktail as an intoxicating foam and meanders through the rest of the cocktail as a web of complex and yet pleasing strings. The whites impart no flavor, but by thickening the drink they also detract from its natural refreshment quotient. This is why I suggest the fizz as a wintertime sour.

1/2 ounce fresh lemon juice
1 large egg white, lightly beaten
2 ounces gin, whiskey (any kind), or rum
1/2 ounce Simple Syrup (page 24)
2 dashes of angostura bitters
Splash of club soda
Splash of red wine (optional)

Put the egg white and lemon juice in a shaker and shake vigorously 10 times. Add the whiskey, simple syrup, and bitters, and fill the shaker with ice. Shake 20 times.

Strain into a rocks glass with no ice and top with a splash of soda. A splash of red wine over the mousse is tasty.

NOTE: It won't surprise anyone that I recommend fresh, organic eggs for fizzes, but if you are concerned about using raw egg whites, consider this: you are rapidly shaking them in a bath of lemon and alcohol, which sterilizes them. You can also use pasteurized egg whites, if you must.

CLOUDY PEAR

Serves 1

This is not exactly a classic sour, because it allows the addition of a dash of other liquors, so that it is halfway between this chapter and the next, but it does introduce the technique of complicating a drink by adding a tiny dash of a secondary liqueur. Other liquors you might want to consider rinsing, or coating, a glass with for sours are Chartreuse, Pernod, and red wine. Coating is a way to control a small amount of potent liquor and keep from overusing it.

2 ounces pear-infused vodka (see Note)
or store-bought pear vodka
1/2 ounce fresh lemon juice
1/2 ounce Simple Syrup (page 24)
Poire Williams eau-de-vie
Sambuca
Splash of club soda
Dash of angostura bitters

Pour the vodka, lemon juice, and simple syrup into a cocktail shaker. Cover and shake vigorously 10 times. Remove the cover, fill the shaker with ice, cover, and shake vigorously 20 times.

Pour a little eau-de-vie and sambuca into a rocks glass and swirl them around in the glass so that they coat the bottom and the sides. Pour any excess from the glass.

Fill the glass with fresh ice and strain the cocktail over the ice. Add a splash of club soda and then dot the surface of the drink with bitters. Serve.

NOTE: To make pear-infused vodka, put 2 well-washed ripe Bartlett pears in a 2-quart lidded jar. Add the peel of 1 lemon and 1 liter of vodka. Cover tightly and set aside for 2 days. Strain the vodka and return the strained liquid to the original vodka bottle. This will keep as long as any other vodka.

FLAVORED SYRUPS

Many home bartenders like to dress up simple sours with flavored syrups, and indeed they add seasonal flair and drama to these straightforward cocktails. Professional bartenders often make complicated syrups, such as cranberry-pine-ginger or grapefruit-chamomile, to riff on simple drinks. I don't suggest you go this far, but an easy syrup is a pleasant way to add interest to a sour. Make these well ahead of time; you will find numerous ways to use one or more in many cocktails. The rosemary syrup is great in a classic gin or vodka sour or as a sweetener for a white sangria. The spiced cranberry syrup works well in a whiskey sour or a dark rum daiquiri.

ROSEMARY SYRUP

Makes about 2 cups

2 cups water
5 long (5- to 7-inch) sprigs rosemary
2 cups sugar

In a large saucepan, bring the water and rosemary to a boil over medium-high heat. Boil for 5 minutes. Add the sugar and cook, stirring now and then, until the sugar dissolves, about 2 to 3 minutes.

Strain the syrup, discard the rosemary, and let the syrup cool to room temperature.

You can transfer the syrup to a lidded container and store in the refrigerator for up to 2 weeks.

SPICED CRANBERRY SYRUP

Makes about 8 cups

4 cups sugar
2 (1-pound) packages fresh cranberries
5 (3-inch) cinnamon sticks
1 tablespoon whole star anise
4 cups orange juice

In a sturdy pot, mix the sugar, cranberries, cinnamon sticks, and star anise. Add just enough water to make a thick paste and bring to a boil over medium-high heat, without stirring. Immediately add the orange juice and stir well, crushing the cranberries with the back of a wooden spoon.

Return the liquid to a boil, remove from the heat, and strain into a bowl through a fine-mesh sieve. Set aside to cool to room temperature.

You can transfer the syrup to a lidded container and store in the refrigerator for up to 2 weeks.

CHAPTER 5

THE COMPLEX SOURS

Sidecar
Aviation
Dos Caminos Margarita
Paradiso
Ruby Foo

Makeup: Primary spirit and flavoring spirit plus acid

Complexity: Low to high

Sweetness: Medium to high

Acidity: High

Strength: Medium

Level of Refreshment: Medium

With all due respect to the evangelists of the martini and to the revisionist camp of the mint julep, complex sours comprise the most important family of cocktails, from both a historical and a contemporary perspective. No other category is as capable of absorbing new ingredients—found or created—and making them work together as this one. Indeed, credit may be given to it for reviving our cocktail instincts in general.

THE INTERNATIONALISTS

Some people may dismiss complex sours as too sweet, too fruity, or too girly, but to make any of these claims is to overlook how adaptable these drinks are to different cuisines. Their ability to accommodate fruits has been responsible for introducing drinkers everywhere to the world of quality cocktails. As silly as these drinks can become, the rules for making a good, balanced cocktail hold as true here as they do anywhere else. And because the architecture of these cocktails is essentially the same as for simple sours, complex sours can be used to lure the novice drinker toward more sophisticated combinations.

I wish I could claim a time delay between the invention of the simple sour and that of the complex sour so that there would be logic to the evolution of the simple to the complex, but alas, it is not clear cut. A review of the last chapter certainly suggests finite permutations of the simple sour, but as the concept of the cocktail gained more and more traction over the last half of the nineteenth century, appearing in more spots across the United States and Europe, the basic language of the cocktail became exposed to an increasing number ingredients. Some of these on their own were no good or were too expensive for mixing but as an additive to a sour lent it waves of flavor. Cocktail books published between 1890 and 1940 were essentially recipes of these drinks duking it out with martini variations for copy space. The margarita, a child of the late thirties, solidifies the complex sour's cocktail world domination, but in truth the concept had been developed years before.

It is not unreasonable to think of this family as a marriage between the tangy simplicity of the simple sour and the liquor-driven complexity of the all-spirits cocktail. The impact of the primary spirit (gin, vodka, whiskey, etc.) is compromised and impacted by a lesser amount of a heavily flavored liquor, one that is occasionally lower in alcohol and almost always sweeter. In this case, the primary spirit extends the usage of the secondary spirit, which would be too concentrated to consume on its own. By the same token, the secondary spirit connects the primary spirit to the juice, enabling them to blend together. The fruit juices keep these drinks from becoming as nuanced as the all-spirits cocktails, and the combination of three or more ingredients can lead to some unexpected and alluring flavors.

There is no difference between how you make a complex sour and how you mix a simple one; all the same principles apply. Shake long and hard and strain over fresh ice if you are serving on the rocks.

SIDECAR

Serves 1

This is a noble drink. While not the first of its kind, it set the paradigm for how a cocktail was constructed in the twentieth century. When I have a sidecar, I am always amazed by the emergent flavors of cocoa and stone fruit that derive from the combination of the two liquors, accented by the citrus. It is a subtle drink that no one should feel ashamed to consume.

1 lemon wedge
2 to 3 tablespoons granulated sugar
2 ounces cognac
1/2 ounce Cointreau
1/2 ounce fresh lemon juice

Rub the rim of a martini glass with the lemon wedge. Spread the sugar on a small plate large enough for the rim of the glass. The sugar should be about 1/4 inch deep. Dip the rim of the glass into the sugar to coat. Set the glass aside.

Pour the cognac, Cointreau, and lemon juice into a cocktail shaker and fill the shaker with ice. Cover and shake vigorously 20 times. Strain into the sugar-rimmed glass.

AVIATION

Serves 1

Every year, it seems as if a long-forgotten complex sour is "rediscovered" by a studious bartender who has been poring over out-of-print cocktail books. Once it shows up on a menu, it spreads like wildfire throughout the professional community as the next "big thing." The aviation was the first of these to illustrate the depth of this family of cocktails, and in some ways it demonstrates how just three or four ingredients can produce a libation of surprising complexity in a world where cosmopolitans once ruled. If you like this cocktail, seek out others, such as Twentieth Century or The Last Word, two cocktails from the same era.

2 ounces gin
1/2 ounce maraschino liqueur
1/2 ounce fresh lemon juice
Crème de violette (optional)
Cocktail Cherry (page 25), for garnish

Pour the gin, maraschino liqueur, lemon juice, and a splash of crème de violette into a cocktail shaker. Fill the shaker with ice. Cover and shake vigorously 20 times.

Strain the cocktail into a martini glass and serve, garnished with the cherry.

DOS CAMINOS MARGARITA

Serves 1

The Dos Caminos Margarita pays my salary. We sell more margaritas than anything. When B.R. Guest Restaurants first opened the Mexican restaurant Dos Caminos, we concocted dozens of variations on this warhorse and found that this one was the hands-down favorite. This led us to two discoveries. First, Triple Sec makes a better secondary spirit in a margarita than do higher-end liqueurs such as Grand Marnier and Cointreau. I think if you use a quality tequila—something that was nearly unthinkable even ten years ago but is quite possible now—a "better" secondary liquor is apt to overpower the primary spirit and mask the herb and fruit aromatics that make tequila one of the world's great spirits. Neither Grand Marnier nor Cointreau works as well with lime as with lemon, either, and the relative blandness of the Triple Sec seems to be a strength. Our second discovery was that a bit of lemon juice, certainly not traditional, seems to brighten up the drink.

Although purists would rather a cocktail be called a margarita only when it's made with lime juice, the cocktail remains unmatched as a drink for absorbing other fruit flavors. Strawberry, passion fruit, and fresh white peach are some of the stand-in margaritas we have served at Dos Caminos restaurant with great success. Simply replace the lemon juice and its proportions with a fresh puree of your favorite seasonal fruit.

1 lime wedge
Kosher salt
1 1/2 ounces 100 percent blue agave blanco tequila
1/2 ounce Triple Sec
1/2 ounce fresh lime juice
1/4 ounce fresh lemon juice
1/4 ounce Simple Syrup (page 24)

Rub the rim of a rocks glass with the lime wedge. Spread the salt on a small plate large enough for the rim of the glass. The salt should be about 1/4 inch deep. Dip the rim of the glass into the salt to coat. Do not oversalt. Set the glass aside.

Pour the tequila, Triple Sec, lime juice, lemon juice, and simple syrup into a cocktail shaker. Fill the shaker with ice, cover, and shake vigorously 20 times. Fill the rocks glass with ice, strain the drink over the ice, and serve.

PARADISO

Serves 1

I invented this drink for the restaurant Fiamma, to see how many levels of sympathetic flavors I could establish in one glass: the amaretto and the peach flavors work together, as do the rum and lemon, which keep everything from becoming too sweet. The white pepper? I was bored and thought it was a decent pun on the white peach. It turned out to be what holds the drink together; the little afterburn of spice keeps the fruit from taking over.

WHITE PEACH–WHITE PEPPER FOAM

2 gelatin sheets
2 cups white peach puree (see Note)
2 cups limoncello
2 teaspoons freshly ground white pepper

COCKTAIL

1 1/2 ounces golden rum
1/2 ounce amaretto
1 ounce fresh lemon juice

To make the foam: In a small bowl filled with ice water, soak the gelatin for about 15 minutes to soften.

In a saucepan, mix together the peach puree, limoncello, and pepper and bring to a full but gentle simmer. Remove the pan from the heat. Lift the gelatin sheets from the ice water, gently squeeze out the water, add the sheets to the saucepan, and let the gelatin dissolve. Set aside to cool slightly.

Put the peach mixture into a chilled iSi Gourmet Whip or Cream Whipper canister or similar foamer. Refrigerate for at least 1 hour.

To make the cocktail: Pour the rum, amaretto, and lemon juice into a cocktail shaker. Fill the shaker with ice, cover, and shake vigorously 20 times. Fill a rocks glass with fresh ice and then strain the cocktail into the glass.

Charge the chilled foam canister with 2 N_2O cream chargers and shake 5 times. Use immediately to top the cocktail with foam.

NOTE: Although you can buy frozen peach puree in some specialty markets, you can also make it. Peel and pit ripe peaches and puree them in a food processor fitted with the metal blade or a powerful blender. For 2 cups of peach puree, you will need at least 8 to 10 peaches and perhaps more, depending on their size and degree of ripeness. If you cannot find white peaches, use yellow peaches.

RUBY FOO

Serves 1

This enduringly popular cocktail's namesake is the Asian restaurant Ruby Foo's, where we pour it by the bucket load. The intent was to meld the sensibility of the cosmopolitan with Eastern flavors, and the flavored sake, of course, plays the role of the flavoring liquor. While it's not something I invented, when I look at its popularity, I wish I had!

1 ounce vodka
1 ounce plum sake
1/2 ounce cranberry juice cocktail
1/2 ounce pineapple juice
1 lemon round, for garnish

Pour the vodka, sake, cranberry juice, and pineapple juice into a cocktail shaker. Fill the shaker with ice. Cover and shake vigorously 20 times.

Strain the cocktail into a martini glass and serve, garnished with the lemon round.

CHAPTER 6
MUDDLED DRINKS

Mojito
Wildwood Mint Julep
Black Cherry Bounce

Makeup: Spirits and solids

Complexity: Medium

Sweetness: Variable

Acidity: Variable

Strength: High

Level of refreshment: Variable

In the preceding chapters, we have traveled the cocktail's evolutionary path, from the simple to the complex. Along the way we gradually added techniques and ingredients. In this chapter, which I would call "The Salads" if I had been compelled toward clarity in my writing, we depart from that trajectory to a family of drinks that exist outside the paradigm. The drinks in this group are not related by ingredient or by history, but by technique. Muddling results in cocktails that are completely different from anything we have described thus far and allows us to construct drinks that are entirely singular and distinctive. On page 3, I describe how to construct a makeshift bar to support what we've been doing so far. The same setup will work—to some degree—with muddled drinks, although they have a testy habit of throwing a wrench into the works.

WHAT IS MUDDLING?

Muddling is the best way to blend solids, such as fruits, herbs, and sometimes spices, into a drink. To do so, in a dry or semidry cocktail shaker, a bartender crushes a substrate to release its juices, oils, and aromatics, which will then blend with alcohol. There is no other way to get such fresh ingredients into a drink. This is the strength of this group of cocktails. They blow our minds and slake our thirst with the best the season has to offer. But its strength betrays its difficulties: muddling is time consuming, potentially messy, and does not alone create a mixed drink.

The technique comes from a lack of bartending tools: shakers and good-quality ice. To understand what I mean, assume that mixing drinks happened simultaneously with the development of bartending tools. Before the advent of the ice machine, bartenders, especially those setting up shop in warmer climes, did not have access to large clean block ice. They may have had access only to shaved or crushed ice, which melts if you try to shake with it. Unable to mix drinks in the ways I have described in previous chapters, the bartenders developed muddling as a dry, ice-free way of mixing drinks.

A CURIOUS RENAISSANCE

When I was coming up as a bartender, muddling had largely fallen by the wayside in most mainstream bars, its last link to contemporary bartending an oft-lost and rotting red muddler idling in the supply drawer behind the bar.

Then along came the mojito, and everything changed. Suddenly, every bar that thought itself worthy, whether it was or not, had to make mojitos. Fresh mint appeared in wilting bunches at the service end of every station. Seasonal variants and extra fruits to throw into the shaker appeared, too, pushing the drink farther from the world of cocktails and closer to fruit salad. I sometimes think that every ingredient that can't be incorporated into a traditional cocktail finds its way into a muddled one. It is for this reason that muddling is the friend of the white spirits, especially vodka, because the flavors we add to it are so strong and vibrant that the spirit can simply blend into the background, which it does very well.

As anyone who has had to dodge the arched eyebrow of a busy bartender knows, muddled drinks are extraordinarily time consuming. Plus, making the drink, with the different mechanics and dirtier equipment needed, interrupts the basic flow of professional bartending. Don't get me wrong. I do not dislike these drinks; I simply feel that home bartenders should know what they are getting into when deciding to make a muddled drink.

MUDDLING AT HOME

When someone tells me that he or she is having a party and plans to make mojitos, I pray that I will not be invited. And if I am invited, I pray that I am so distracted by scintillating conversation that I will not be tempted to enter the kitchen and see its attendant mess of mint and lime on the countertop, the melted ice on the floor—and on the host's clothing. Bearing the onus of keeping up with the guests' drinks, the poor party giver, now reduced to a sweaty masher of things, will not have a chance to leave the kitchen and have a laugh or two. If I see this, professional courtesy compels me to take pity, put aside my own cocktail, and offer assistance—at which point any chance I have to reenter the party for the next few hours is probably shot.

On the other hand, to welcome a few good friends to a barbecue or to celebrate the end of a perfect summer's day, there is no drink that is as joyful and refreshing as a cocktail muddled up with seasonal ingredients. Be aware that you must be able to control how many of these drinks you have to make, and be sure you have a good setup prepared. Before you begin, have plenty of cut fruit in bowls and keep herbs refreshed in water.

When you decide to try your hand at muddling, purchase a proper muddler. Any good cooking store sells them. Rolling pins, bar spoons, and the spout ends of bottles do not work. Believe me: I have witnessed horrible things committed in the name of muddling.

Muddling is easy to do. Hold the container in which you are muddling (pint glasses remain the best for this) securely on a flat surface. Firmly crush the ingredients with the muddler, using a slight rocking motion. Muddling does not have to be quick or fierce. Break apart the ingredients, but do not obliterate them—ingredients, especially herbs, can be overmuddled as easily as undermuddled.

In general, most muddled drinks should be served over crushed ice. This makes them truly, deeply, refreshing and gives them the glowing snow cone look that makes them so appealing. You can buy linen bags (Lewis bags) for ice cubes that can be whacked with the muddler at any store with a good bartending selection, but bear in mind once again, crushing ice is time consuming. Deputize a friend of yours as a barback or, even better, prevail on friends whose freezers have the crushed ice option to entertain more. And have them make the drinks.

MOJITO

Serves 1

I don't believe in taking shortcuts with mojitos. I have had too many undrinkable ones to think otherwise. The time saved has never been worth the absence of pleasure.

I admit that I am very particular. For instance, you will note that I call for mint sprigs, although some bartenders like to use the leaves only. I prefer the stem's stronger herb oils and like the fact that the stem keeps the leaves fresh while the drink is being consumed. At least half the joy in a mint cocktail is in the aroma. Thus, I endeavor to force the imbiber's nose into a thicket of fresh mint leaf garnish.

I train bartenders to go through a step-by-step process when learning how to make a mojito, so that each technique makes up a perfect whole. Only by practice do they—and will you—learn how to make this gorgeously refreshing cocktail, and so I have departed from my usual way of writing a recipe. Walk with me.

1. To the glass or bottom half of a cocktail shaker, in this order, add:

3 bar spoons or teaspoons granulated white
or demerara sugar
3 roughly torn mint sprigs
3 lime wedges

2. Muddle firmly 5 or 6 times. What is in the shaker is a "sandwich" for maximum muddle efficiency. The sugar on the bottom acts as an abrasive and helps rip the mint leaves apart. The lime wedges on top absorb the full force of the muddle and release juice even while protecting the mint from being bruised. You don't want to rip the mint apart at this point; just weaken it. Never overmuddle mint. It ends up losing its vivaciousness and tasting like overcooked greens.

3. Add:

2 1/2 ounces white rum

4. Fill the shaker with ice and shake 10 times, at most. You do not need my ordained 20 shakes, because we've already blended most of the ingredients. At this point, you need only to stun the mint into falling apart. The mint is already weakened from muddling, and so it does not take a lot of shaking to do the job.

5. Fill a highball glass with crushed ice and strain the drink over it. If you prefer a cocktail with more evidence of mint and lime, simply pour the contents of the shaker, ice and all, into an empty glass.

6. If you like, top the drink with a splash of:

Club soda or citrus soda

7. Garnish the drink liberally with:

Attractive mint leaves

(Keep in mind that mint smells stronger than it tastes, so be as generous with the mint as you can. The bouquet will trick the brain into tasting
it more fully.)

8. Nestle a straw in their midst and serve.

WILDWOOD MINT JULEP

Serves 1

Historically, the mint julep is the most time consuming of all drinks to make. At Wildwood, a Manhattan barbecue restaurant, the challenge I faced was developing a julep that the bartender could make without having to spend the usual time. At a busy New York bar, no barkeep has time to make a julep the way it was meant to be made on a sultry southern evening with a soft breeze wafting through the wisteria. The process I developed ended up exposing how this drink works. While I am sure you will make it in a rocks glass, if you happen to have a pewter julep cup, chill it in the freezer and use it for your cocktail. And crushed ice is not optional. It's necessary for the julep to work.

3 sprigs mint
1/2 ounce Simple Syrup (page 24)
2 ounces full-bodied bourbon or cognac
Fresh mint leaves, for garnish

In a rocks glass or chilled julep cup, combine the mint and syrup and muddle gently. Do not break the mint apart. Instead, crush it against the sides of the glass. Let the mixture sit for at least

1 minute, although 5 or even 10 minutes is better. (This is the perfect time to make another drink or clean up the bar area.)

Add the bourbon and stir lightly, just to mix.

Fill the glass with crushed ice and gently swizzle the drink. Mound a little more crushed ice on top.

Garnish liberally with mint leaves and nestle a straw in their midst.

The Best Mint

What mint makes the best mint julep? At the opening of Wildwood, a barbecue restaurant, I assembled eight different types of mint, among them peppermint, spearmint, Havana, and pineapple. I conducted an experiment with different mints used to make a number of juleps and eliminated them by taste, play-off style. While none of the mints made a bad drink, the cocktails were amazingly different. The winner was a mint called, appropriately enough, Kentucky Colonel. It exhibited all the fresh summery joy of the herb but did not overpower the bourbon. It is not hard to find and, like all mint, is easy to grow. If you live in a temperate climate, it will grow like a weed.

BLACK CHERRY BOUNCE

Serves 1

This nontraditional muddled drink is great in the summer and it is extremely popular at the restaurants Blue Water Grill and Ocean Grill. The maraschino liqueur underscores the black cherries and bolsters the flavor of the fruit. The same principle could be used with any combination of fruit and spirits—for example, orange liqueur with fresh citrus or crème de pêche with peaches or nectarines.

3 fresh black sweet cherries, pitted and stemmed
3 thin lemon rounds
1/2 ounce Simple Syrup (page 24)
2 ounces London-style dry gin
1/4 ounce maraschino liqueur
Angostura bitters
Club soda
Black cherry, for garnish

In the bottom of a cocktail shaker, gently muddle the cherries, lemon rounds, and syrup so that the cherries break but do not fall apart.

Add the gin, maraschino liqueur, and a splash of bitters. Pour into a water glass filled with fresh ice. Top with a splash of soda and garnish with a black cherry.

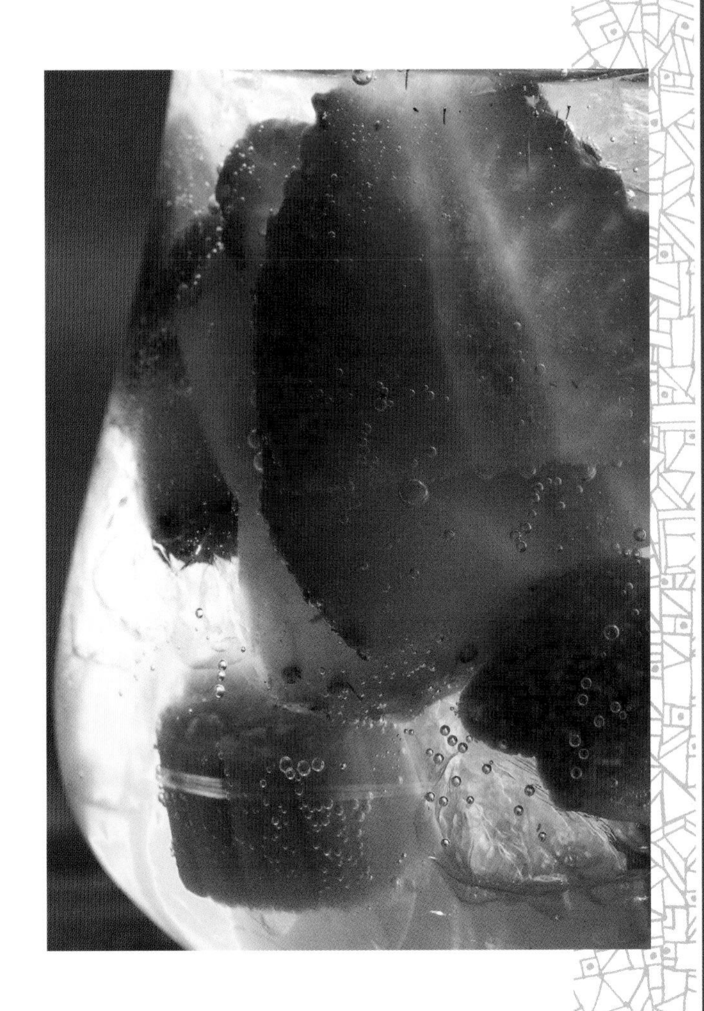

CHAPTER 7
HIGHBALLS

Perfect Harvest
Persa
Watermelon Paloma

Makeup: Mixers are the majority

Complexity: Variable

Sweetness: Variable

Acidity: Medium to low

Strength: Low

Level of Refreshment: High

Highballs can be as simple or as complicated as we want them to be. They lean toward the former, because most are so simple to make. The simplest are nearly infallible. They're what to order when you are not sure you want a cocktail. And because the primary ingredients are nonalcoholic liquids, they're what to enjoy when you don't want a full-blown cocktail. It is even a bit of a stretch to call them cocktails because none of the techniques we've applied before really fit here. Far better to bump them just slightly down the evolutionary ladder and refer to them as mixed drinks.

Highballs and beer, of course, are the bread and butter of a busy bar, the beverages we can pump out rapidly to give us the time to prepare the more time-consuming drinks. You can use this to your advantage during your home event, too. Having a pitcher or two of a mixer so that the guests can fix their own takes some of the pressure off of you.

A classic highball should be considered a tall drink in which a spirit or two is diluted with soda or juice. The vodka and tonic, the screwdriver, and the bloody mary are all examples of this. Because water is the primary ingredient, there is no need to shake with ice before serving. There's enough liquid in there already. Fill a highball glass (a tall 12-ounce glass) with ice, pour in 1 1/2 ounces of liquor, and fill the rest with the mixer. Stir with a straw if you want. For thicker highballs such as bloody marys, you might want to pour the drink once or twice between two glasses to blend in the liquor; but in soda-based drinks, the effervescence will do the work for you.

A modern highball can be a little more labor intensive. This is the term I give to tall drinks that might have multiple liquors and mixers but still are consumed as classic highballs. In this case, some of the principles and techniques of complex sours apply. Whether simple or more complex, if you want to make a light-bodied, seasonal drink, you'd be hard pressed to think of any other than a highball.

PERFECT HARVEST

Serves 1

Highballs underscore my conviction that this is the best group of cocktails for taking advantage of the bounties of the season. The spirit components can be intellectual and complex and still remain thirst quenching, as evidenced in this perfect autumn cocktail.

1 1/2 ounces bourbon
1/2 ounce Meletti Amaro
1 ounce fresh apple cider
1/2 ounce Spiced Honey (recipe follows)
Club soda
1 thin apple slice

Pour the bourbon, amaro, cider, and honey into a cocktail shaker. Fill the shaker with ice. Cover and shake vigorously 20 times.

Fill a highball glass with ice and strain the cocktail into it. Top with a splash of soda and garnish with the apple slice.

SPICED HONEY

Makes about 16 ounces or 2 cups

1 cup honey
1 cup water
1 teaspoon freshly grated nutmeg
1 teaspoon ground coriander
1 teaspoon ground cinnamon

In a saucepan, combine the honey and water and bring to a simmer over medium heat. Add the nutmeg, coriander, and cinnamon and cook, stirring, until the honey dissolves.

Strain the honey through a fine-mesh sieve. Let the honey cool and then store in a tightly lidded jar in the refrigerator for up to 1 month.

PERSA

Serves 1

I created a menu of cocktails for Italian restaurant Vento. After coming up with cocktail name upon cocktail name, each one translated into Italian. I was at a complete loss for yet one more. So I used the word *persa*, which means "lost." It's a lovely-sounding word to describe a gorgeous springtime drink, which combines two of spring's best flavors: rhubarb and strawberry. One of the best things about infusing your own vodka is that you can capture a season and in doing so create a cocktail that is unique to the time and place when and where it's made.

We know spring is here and summer is around the corner when guests start checking the cocktail menu to see if we have added the Persa yet. They love it but understand it's offered only during the warm months because of the seasonality of the infused ingredients. The rhubarb infusion requires three days to macerate, and so when you want to make this drink, plan ahead! It's well worth it, and your guests will welcome the opportunity to greet the warm weather with a Persa toast.

1 1/2 ounces Rhubarb Infusion (see page 110)
1 ounce moscato d'asti
1/2 ounce Triple Sec
1/2 ounce Simple Syrup (page 24)
1/2 ounce fresh lemon juice
2 to 3 strawberry slices, for garnish

Fill a highball glass with ice. Pour the rhubarb infusion, moscato d'asti, and Triple Sec over the ice. Add the syrup and lemon juice, garnish with strawberry slices, stir briefly with a straw, and serve.

WATERMELON PALOMA

Serves 1

The paloma is a classic Mexican highball made with tequila and high-quality grapefruit soda, which is a popular flavor in Mexico. At Dos Caminos, we use this as a template for adding fresh ingredients. Try this also with clementine or guava juice. Sipping one of these is truly the best way to watch the sun set on hot summer's day.

1 lime wedge
Mint Salt (see page 110)
1 1/2 ounces blanco tequila
2 ounces Jarritos grapefruit soda or Squirt
1 ounce Watermelon Fresca (see page 111)

Rub the rim of a highball glass with the lime wedge. Spread the salt on a small plate large enough for the rim of the glass. The salt should be about 1/4 inch deep. Dip the rim of the glass into the salt to coat.

Carefully fill the glass with ice. Pour the tequila into the glass. Add the grapefruit soda and fresca and serve.

MINT SALT

Makes about 1 cup

1 cup kosher salt
2 sprigs mint

In the bowl of a food processor fitted with the metal blade, process the salt for about 30 seconds to grind it finely. Add the mint sprigs through the feeding tube and process until well mixed with the salt.

Transfer the salt to a lidded container and store at room temperature for up to 1 month.

NOTE: To make a smaller amount of mint salt, you can use a mini food processor, a coffee grinder, or a spice grinder.

RHUBARB INFUSION

Makes about 1 liter

1 stalk fresh rhubarb, leaves removed
6 strawberries, hulled
1 liter vodka

Chop the rhubarb into pieces about ½ inch wide. Cut the strawberries into thin slices. Transfer the fruit to a glass, ceramic, or other nonreactive container.

Pour the vodka over the fruit. Stir gently, cover, and refrigerate for 3 days.

Strain the fruit from the vodka. The strained vodka will keep indefinitely.

WATERMELON FRESCA

Makes about 4 cups

4 cups rindless 2-inch watermelon chunks

2 cups fresh orange juice

Put the watermelon chunks into the bowl of a food processor fitted with the metal blade. Add the orange juice and puree until smooth.

Strain through a fine-mesh sieve. You can refrigerate it in a tightly lidded container for up to 2 days.

NOTE: This makes a lot of fresca, but it's worth it. You can also drink it for breakfast on its own!

ACKNOWLEDGMENTS

The best bar books never written all end up as cigarette smoke, blown toward the ceiling fans of after-hours bars around the world, in the postshift conversations of bartenders relieving their stress with equal parts booze and equal parts war stories. This book is written in acknowledgment of all those bartenders and the people who follow, who kept this book from becoming like the ones just described.

I usually tell people it was just good timing for me to have been bartending in this resurgent age of cocktails, but my whole life has been a bit of a cocktail age—from my parents', Kai and Bonita's, early goofy stabs at home brewing and winemaking to their successes later; from Grandpa Ram's letting me sample his gimlets, probably meant to quiet me while he watched golf; to Luke Johnson's and my early attempts, pompous Ivy League style, to understand the nuances of the martini. And memories (vague, at best) of the Dobsonfly, Don't Say Please Say Gimme, and the Sky Bar Is Open!!!!

To those from my formative years, without your temptations I might have remained in science: Karla, who at Le Zinc served me my first Negroni; the ladies of 503, who let me invent for them; Mo; Ross, who saved my life at five and then destroyed it at twenty-eight by giving me my first bartending job. You created a monster.

Those first years in New York were wonderful and terrible, as I learned that there were others abnormal enough to be thinking deeply and forwardly about the state of the American mixed drink. I have never learned more from a bartender than from watching Chris Lynch say thank you to a guest. Adil, Rachid, and John, who gave me intellectual freedom in return for learning great service. In the past decade, I've had the opportunity to talk and share drinks with, by, and from every bartender I've ever admired or wanted to meet. The things you are doing inspire me to do better. There are not enough pages allotted to name you all or recall what we drank, but if it was great, I remember it. If it was bad, I remember that, too. One great moment will have to suffice: martinis stirred by Suzi at the Hotel Adlon with my best friend, while the sky turned slowly into the color of our oysters.

And mostly, to the carnival that is B.R. Guest. Thanks to Steve Hanson, whose continuous quest to make his restaurants better leads him to such extreme ends he does crazy things like hire me, and never lets me rest on my successes. Without him, this book, and 500 or so of my last good ideas would never have existed, as well as 127 of my bad ones. Donna Rodriguez patiently shepherded this book from a conference call idea into a physical reality I still can't quite believe, especially after her aesthetic judgments made that reality better to look at. Jane Dystel, my agent, who introduced me to Mary Goodbody, who asked all the right questions I needed to answer for this book. The Wine and Spirits Department past—Greg, Tim, Jonill, Jasper, Bill—but especially present, Alex Cauchon and Laura Maniec, whose enthusiasm and zeal for the business of serving drinks rivals my own, and whose creativity and energy makes a lot of what I do possible. Scratch that, all of it. Every day is a treat because of the feedback and support I get from the scores of chefs, managers, and receivers who make our restaurants great, and who help me turn paper ideas into liquid realities. And lastly, to go full circle, to the bartenders, barbacks, and servers who put passion and love into their work every day, and in doing so inspire me to keep their lives interesting.

Eben Klemm did not choose the most typical path for one destined to be a master of creative cocktailing: he managed a molecular biology laboratory at the Whitehead Institute of Biomedical Research at MIT. Born in 1970, Klemm grew up in upstate New York and studied biology at Cornell University. Before MIT, Klemm worked at several biotech companies in the Boston area, but it was in New York City where he would find the next chapter of his life.

Bartending found him for the same reason it does everyone who moves to New York City—which is to say, to simply pay the rent. But trained as he was in the sciences, it was a professional willingness to experiment, as well as a fascination for the culture of cocktails in U.S. history that led him to think about the art beyond the tip bucket. Whatever the impetus, Klemm had the good fortune to arrive at this newly rediscovered profession at a time when the level of creativity at Manhattan bars was rising to match the inventiveness found in restaurant kitchens and on wine lists all across town.

After opening the Campbell Apartment, the retro-cocktail lounge in the landmark Grand Central Terminal, Klemm became bar manager, or the self-applied "cocktaillier," at the three-star Portuguese restaurant Pico in New York's TriBeCa. He was very proud of the program he developed there, but the stress, sleep deprivation, and challenge of replicating it for a large restaurant group was too tempting an offer to pass up. Noted for combinations elegant and playful, daring and addictive, Klemm's cocktails achieved an alchemy all their own and have been featured in such diverse publications as the *New York Times*, the *Wall Street Journal*, *Food & Wine*, *Time Out New York*, *Popular Science*, and *Playboy*. He has also made appearances on MSNBC, CBS's *Early Show*, and ABC's *20/20*.

Beginning with Fiamma Trattoria's opening at the MGM Grand Hotel in Las Vegas in 2003, Klemm has created the cocktail menu and signature drinks for ten restaurant openings and four lounges, as well as revamped the cocktail lists at every B.R. Guest Restaurant.

INDEX